I Can Be Anything!

I CAN BE A PARK RANGER

By Nancy Greenwood

Please visit our website, www.garethstevens.com. For a free color catalog of all our high-quality books, call toll free 1-800-542-2595 or fax 1-877-542-2596.

Library of Congress Cataloging-in-Publication Data
Names: Greenwood, Nancy, author.
Title: I can be a park ranger / Nancy Greenwood.
Description: New York : Gareth Stevens Publishing, [2021] | Series: I can be anything! | Includes index.
Identifiers: LCCN 2019045241 | ISBN 9781538255568 (library binding) | ISBN 9781538255544 (paperback) | ISBN 9781538255551 (6 Pack)
Subjects: LCSH: Park rangers–Vocational guidance–Juvenile literature.
Classification: LCC SB486.V62 G74 2021 | DDC 363.68023–dc23
LC record available at https://lccn.loc.gov/2019045241

First Edition

Published in 2021 by
Gareth Stevens Publishing
111 East 14th Street, Suite 349
New York, NY 10003

Editor: Kate Mikoley
Designer: Laura Bowen

Photo credits: Cover, p. 1 (kid) Sirisak_baokaew/Shutterstock.com; cover, p.1 (background) Lorcel/Shutterstock.com; p. 5 Rob Crandall/Shutterstock.com; p. 7 ArCaLu/Shutterstock.com; pp. 9, 24 Ondrej Prosicky/Shutterstock.com; pp. 11, 24 Aleksandr Stepanchuk/Shutterstock.com; p. 13 Visions of America/Education Images/Universal Images Group Editorial/Getty Images; p. 15 RonTech3000/Shutterstock.com; p. 17 Kevork Djansezian/Staff/Getty Images News/Getty Images; p. 19 Smith Collection/Gado/ Contributor/Archive Photos/Getty Images; p. 21 Jeff Greenberg/Contributor/Universal Images Group/Getty Images; p. 23 LMspencer/Shutterstock.com.

Printed in the United States of America

CPSIA compliance information: Batch #CS20GS: For further information contact Gareth Stevens, New York, New York at 1-800-542-2595.

Contents

Ranger Jon works at a park.
He gives tours.

Don’t feed the animals,
he says.
They can find food.

I saw a big bird!
Ranger Jon says it's
an owl.

Ranger Jon knows about trees too.
This one is an oak!

Rangers make sure we follow rules.
Rules keep us safe.

Bryce Canyon Map and
Bryce Canyon

Parks have rules
about fires.
Ask a ranger before
you make one.

Rangers help put out big fires.

Ann's dad got lost.
Rangers found him!

NATIONA
PARK
SERVICE

Rangers help people.
I like to help too!

10

I can be a park ranger.
So can you!

Words to Know

oak

owl

Index